A Science Museum
Booklet

Barometers

AF886147

A G Thoday BSc ARCS

Her Majesty's
Stationery Office
London

First published 1978

ISBN 0 11 290255 3

Barometers

Introduction and early history

The barometer was invented by Evangelista Torricelli. He was born near Faenza, in Italy, in 1608, the year in which the first telescope was produced. He studied in Rome where he became an accomplished mathematician who did much to develop the subject of hydraulics. He was a great admirer of Galileo, and in 1641 he moved to Florence to take up a position as his secretary. Galileo died a year later and Torricelli succeeded him as Philosopher and Mathematician to the Grand Duke Ferdinand II.

At this time much thought and experiment were being devoted to finding out why it was impossible to raise water out of mines by the then normal siphon method if the water level was deeper than about ten metres (32 ft). This in turn led to a renewed interest in whether or not it was possible to produce a vacuum. Torricelli's contribution was to suggest the experiment that resulted in what we now call the mercury barometer.

In his own description of the experiment Torricelli says he filled two glass tubes with mercury, stopped them with his finger and turned them upside down into a vessel, also containing mercury. The mercury level fell by the same amount in each tube. A drawing of the apparatus shows that one tube had a large bulb at the closed end. The length of the tube was about 86 cm.

Torricelli was convinced that the pressure of the atmosphere maintained the mercury in the tube and that there was a vacuum above the mercury level. In a letter which he wrote to his friend Ricci in Rome on 11 June 1644, it is clear that the first mercury barometer was produced in the first half of 1644 and not 1643 as has generally been accepted. The earlier date originated from an account in *Essays of Natural Experiments made in the Academio del Cimento*, published in 1667, in which it is stated 'that famous Experiment of the Quick-silver is now spread throughout Europe, which first in the year 1643 offered itself to the thoughts of the Ingenious Torricelli'.

The names 'baroscope' and 'barometer' were given to this apparatus by the Hon. Robert Boyle in 1663. Right up to the present day stick and standard barometers vary little in form from Torricelli's first model. The other main type is the siphon-tube variety, which is in the form of a U-tube, with the longer arm sealed off and the shorter arm open to the atmosphere. This was first produced by Robert Hooke in England in 1663, and it was the basis for his wheel barometer which became popular throughout the 19th century. Hooke was the first person to observe and relate changes in the height of the barometer with changes in the weather, shown in a letter from him to Boyle on 6 December 1664.

Although the weather-glass had been in existence for many years before this (see page 36), the earliest example of weather forecasts on the barometer scale is on a barometer made by Daniel Quare in 1680. The first details on how to interpret the changes in the mercury height for foretelling the weather were published in books by George Sinclair of Leith in 1683, *Natural Philosophy improven by New Experiments*, and by John Smith of London in 1688, *A Complete Discourse of the Nature, Use and Right Managing of that Wonderful Instrument the Baroscope or Quick Silver Weather Glass.*

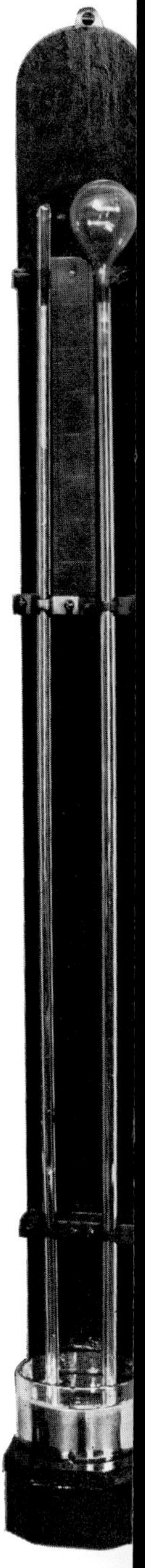

1a *Reproduction of Torricelli's barometer, 1644*

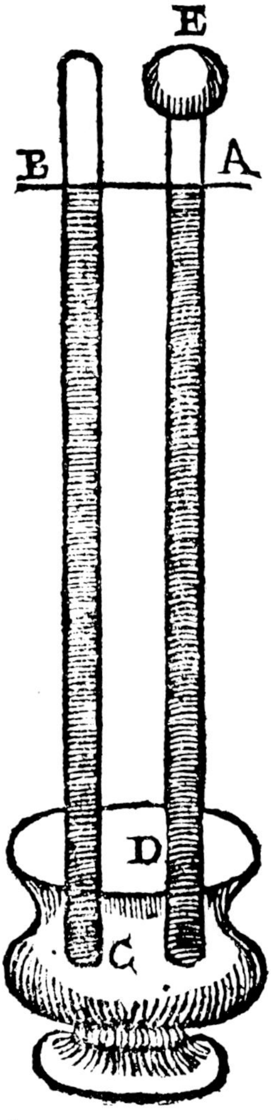

1b *The earliest known drawing of Torricelli's equipment, 1662*

Main types of early barometers

Open cistern types

These were practically identical with the first barometer made by Torricelli, except that the cistern was usually a semi-hemispherical wooden bowl with a similar shaped cover. Examples by J Sisson and J Ramsden, dating from about 1750 and 1770, respectively, are illustrated in Figs 2a and 2b. In the earlier examples allowance was made for the change in level in the cistern with a change of the height in the tube. The cross-section of the bowl was made large relative to the bore of the tube to reduce the error as much as possible. In these examples there is a graduated ivory index dipping into the bowl which enables a correction to be applied to the reading of the height in the tube. In addition, both barometers have a vernier on the scale and, in the former, a reading eyepiece. These are among the earliest barometers to have both refinements to enable more accurate readings to be obtained.

There are comparatively few such barometers in existence today; they appear to have been made for scientific gentlemen who took an interest in recording meteorological phenomena, and were not for general sale.

Closed cistern types

These can be subdivided into those with an entirely wooden cistern and those with a leather base or sides. The closed cistern was made of boxwood in the form of a cylinder into which the tube was cemented. Early in the development of the barometer it was discovered that boxwood was pervious to air, though sometimes there was a hole in the top of the cistern.

Those with the leather base were usually described as portable barometers. There was a flat circular padded plate below this base which could be raised or lowered by a screw, thus pressing the plate into the leather base and reducing the volume of the cistern, with the result that the mercury rose in the cistern and the tube, filling the tube completely. Thus the barometer became portable and could be positioned horizontally. This form is found in the earliest surviving barometers, particularly those by Daniel Quare and the other well-known makers at the beginning of the 18th century. This feature was characteristic of marine barometers up to about 1850, and also of the famous Fortin barometer. Quare patented, against the opposition of other barometer makers in the Clockmakers Company, a portable version in 1695, after first demonstrating it to the Royal Society. Although it has been believed that it was of the leather based type, the wording of the patent gives no indication of the real nature of the form of portability and this must therefore remain an open question.

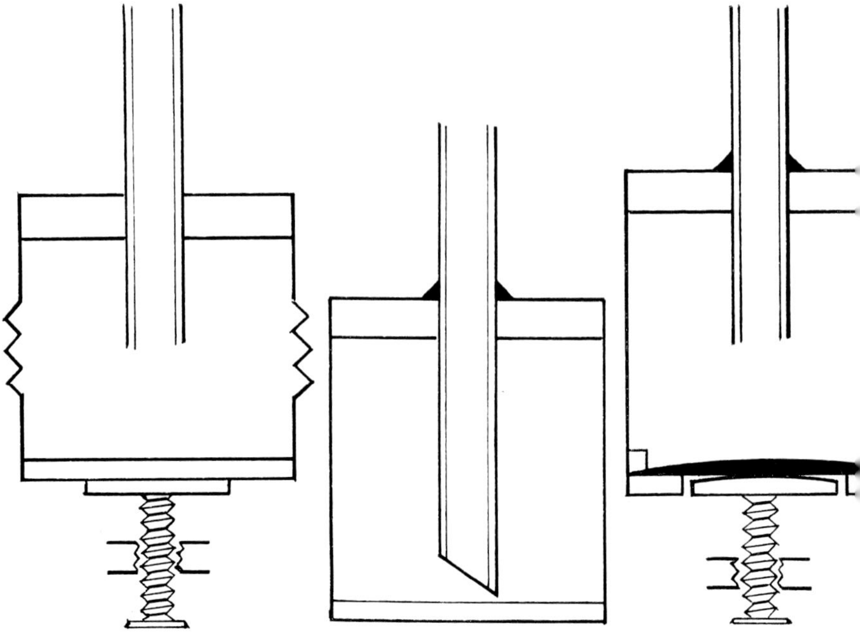

Types of closed cistern c 1646

Siphon types

These were first developed by Blaise Pascal, the great French philosopher, probably about 1646–7. The basic form of the siphon was a U-tube, with the longer limb closed and the shorter limb open to the atmosphere. The difference between the levels of the mercury in the two tubes gave the pressure of the atmosphere, there being a vacuum above the mercury in the closed tube. This simple U-tube form has an obvious disadvantage in that the change in level of the mercury in each limb was half the change of the atmospheric pressure. To overcome this disadvantage Hooke devised a wheel barometer (Fig 3), which he described in *Micrographia* in 1665 (Fig 4).

Early barometers – 18th century

Although quite a number of barometers must have been made in the 17th century, to judge from meteorological accounts in the literature, none made before 1680 seems to have survived. These earliest models were made by members of the Clockmakers' Company such as Daniel Quare, Thomas Tompion and George Graham. Another maker of the same period was John Patrick who was known as the 'Torricellian Operator', but was not admitted to the Clockmakers' Company until late in life. He was probably the principal maker of barometers at the beginning of the

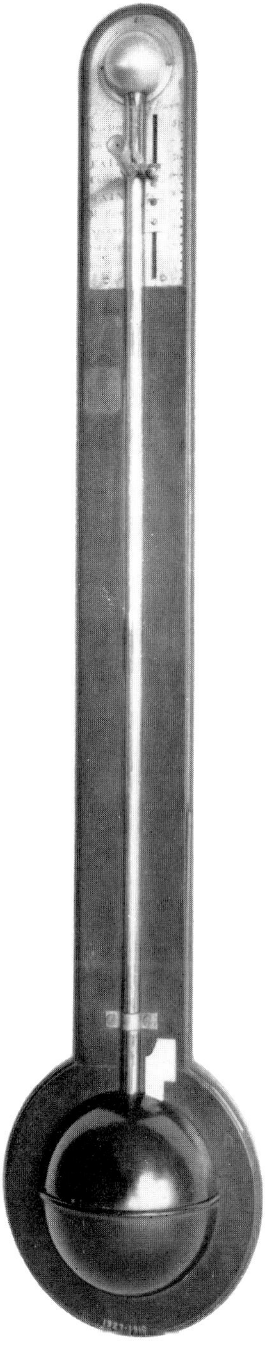

2a *Open cistern barometer by J Sisson c1750*

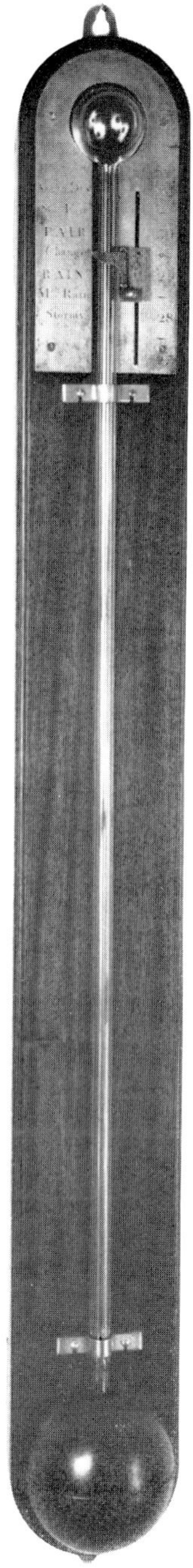

2b *Open cistern barometer by J Ramsden c1774*

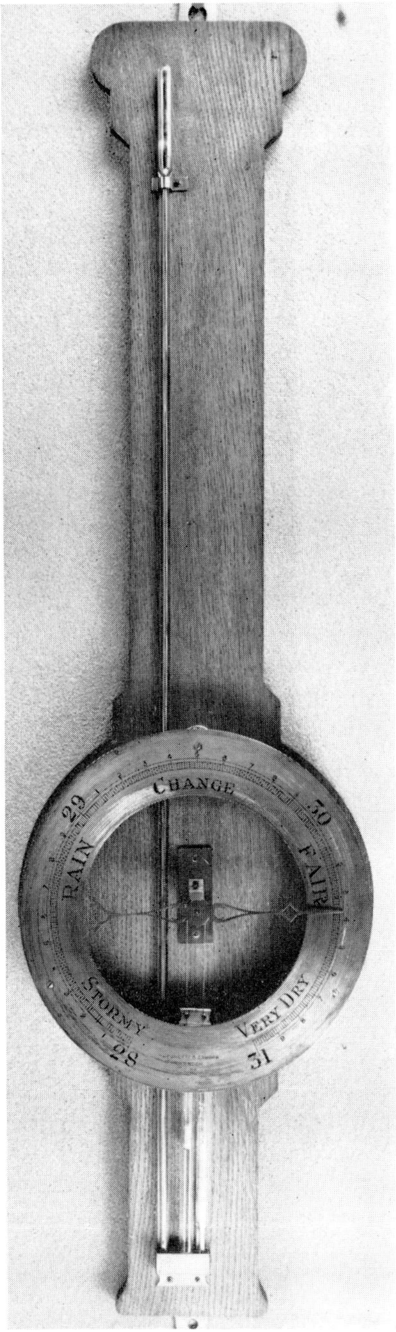

3 *Wheel barometer*

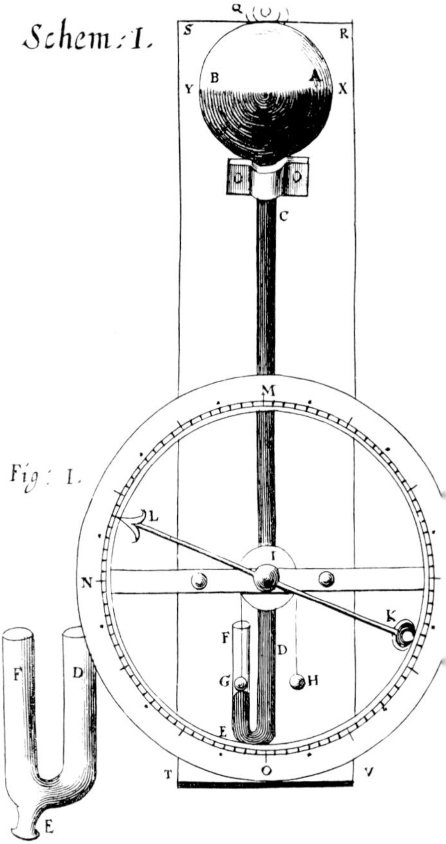

4 *Drawing from 'Micrographia' by Hooke 1665*

18th century, to judge from the appellation he was given and from the number of barometers illustrated on his trade cards. Surprisingly, however, there are very few of his barometers left.

More barometers bearing the name of Daniel Quare survive today than any others of the period. Practically all of them have a characteristic style, (see Figs 5a and 5b), and were mainly of the portable, closed cistern type with a leather base. The two outer finials on the top of his barometers, when rotated, moved the registers up and down to record the pressure. Quare also gave serial numbers to some of his models and these can be found on the right-hand side of the brass case enclosing the scale. On a number of his instruments the weather characteristics are given in English on one side and French on the other. Quare barometers are attractive items of furniture and when they appear on the market command high prices, though fakes are not unknown.

One of the few surviving barometers by George Graham, which dates from about 1720, is notable for its unique arrangement (Fig 6). It is supported on the right-hand pillar, but like many old barometers the tube and cistern are not the original ones. The left-hand pillar supports a spirit thermometer, the scale of which is unusual. The zero of the scale corresponds with the 'Temperate Air' and extends upwards to 80° (Extra Hot) and downwards to 80° (Extra Cold). This was the original scale proposed by Fahrenheit in about 1709 and derived from the Florentine Scale. From the centre of the arch is suspended a cat-gut hygrometer which is held taut by a lead weight below the dial, the catgut being partially enclosed in a tapered walnut wood frame. With change in humidity the gut twists and untwists, and the amount of twist is registered on the dial.

An example of an early siphon type barometer is the one by Isaac Robelou, dated 1719 and mounted on a backboard of Chinese lacquer work (Fig 7). The weather characteristics are given in both Latin and English, with a scale of thirty-six divisions. The scale of the spirit thermometer alongside has another of the early Fahrenheit scales similar to that attached to the Graham barometer above.

The diagonal barometer

The diagonal or 'signpost' barometer was developed to extend the reading on the barometer scale. The tube was bent at a sharp angle at about 70 cm above the level in the cistern, and the upper part of the tube was inclined about 10° or 5° to the horizontal. The movement of the mercury in the top of the tube was thereby expanded over five to ten times, respectively. This form was difficult to read with any accuracy as the mercury meniscus was not at right angles to the side of the tube. The diagonal barometer was invented by Sir Samuel Morland, Master Mechanic to Charles II in about 1680, and an early model was shown to the Philosophical Society at Oxford in 1684.

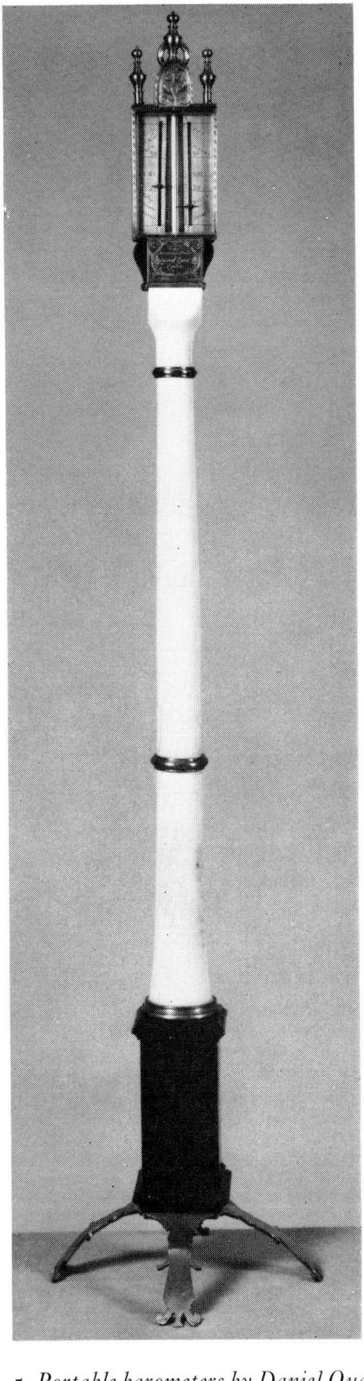

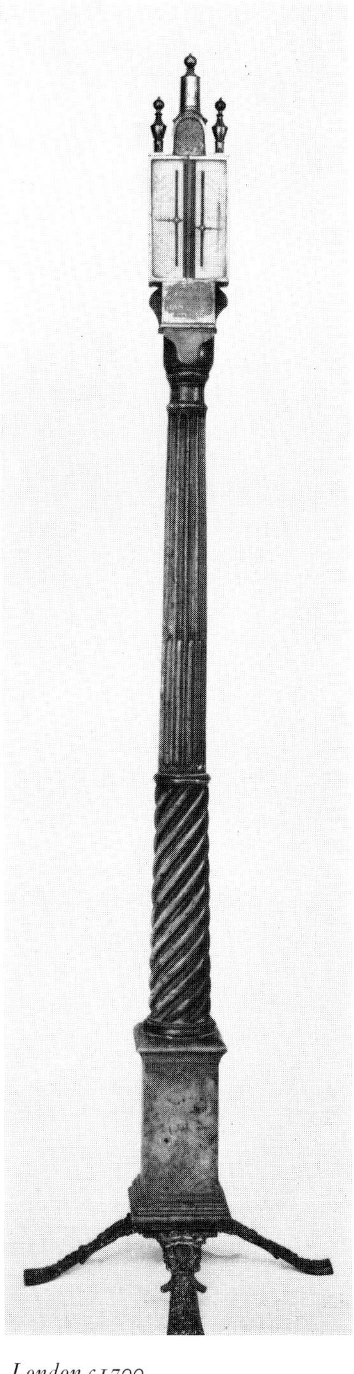

5 *Portable barometers by Daniel Quare, London c1700*
LEFT *in ivory case, Victoria & Albert Museum,* RIGHT *No.38, in walnut case*

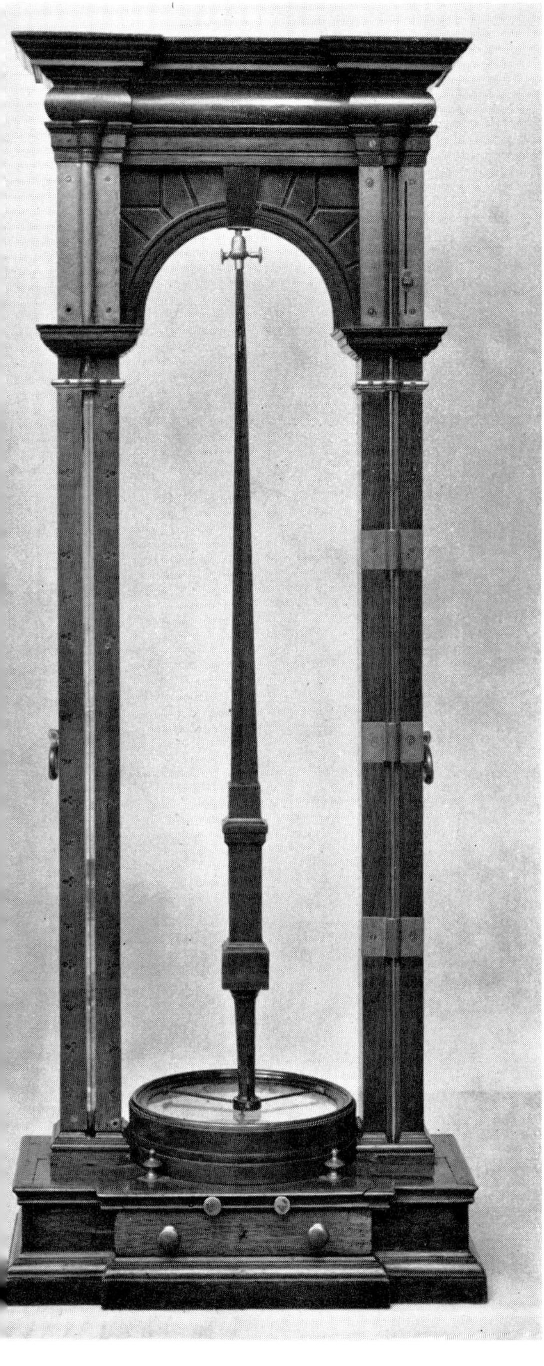

6 *Barometer with thermometer and hygrometer by George Graham, c1720*

7 *Siphon barometer by Isaac Robelou, 1719*

Diagonal barometers were produced mainly during the 18th century. At the beginning of the century John Patrick and Daniel Quare made them and an excellent example by Quare can be seen in Hampton Court Palace. The manner in which Patrick mounted his can be seen in Fig 8, which is an illustration of one of his trade cards. In their simple form diagonal barometers were often ungainly, but Patrick overcame this problem by mounting it around a looking-glass, with a thermometer on

A New Improvement of the Quickſilver Barometer, *being Made* Portable, by *JOHN PATRICK*, in *Ship-Court* in the *Old-Baily* London.

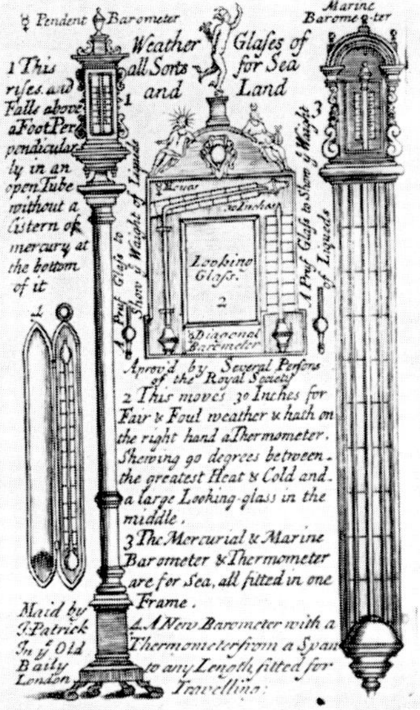

THERE is lately Invented and fitted for Sale a curious Pendant BAROMETER, wherein the Quickſilver riſes and falls (Perpendicularly) in ſome 12, 15, 20, 25 or 30 Inches, inſtead of Three in the Common ones.

It diſcovers the moſt minute and ſmalleſt Alterations in the Air, ſhewing the Weather much ſooner, and more certainly, than any common Barometer; and this in an open Tube (without a Ciſtern of Quickſilver at the bottom;) wherein the expanſion and Contraction of the Column of Mercury is near three Inches, and this in the moſt equal Tube yet met with.

By this the Inclination of the Air is to be known at Pleaſure: For by moving it a little up and down with the hand, you ſhall immediately behold the Quickſilver Riſe and Fall, very conſiderably. Which if it Riſe, it will be Fair Weather; if it Fall, then expect Rain.

It hath alſo been obſerved in ſome of theſe Pendent Barometers, if it be carry'd into a dark Room, and the Quickſilver put into a Vibration, by lifting it up and down, that it will throw forth Flaſhes of Lightning from the top of the Quickſilver, to the top of the Tube in the Vacuum: Which noble diſcovery requires the Solution of the moſt profound Naturaliſt.

This Curioſity has been Examin'd and prov'd by ſeveral Perſons of Quality of the ROYAL SOCIETY, who thought it highly to deſerve the enquiry of the moſt Curious.

ALſo an excellent Diagonal Barometer, wherein the Mercury moves in an oblique Tube, for the ſpace of 30 Inches, inſtead of 3 in the common one; and is ſo Nice as to divide an Inch into 100 Parts. For the ſame purpoſes as the former.

It has a Thermometer on the ſame Frame, ſhewing 90 Degrees of Variation between the greateſt Heat, to the greateſt Cold.

And a Looking-Glaſs commodiouſly plac'd on the ſame Frame, between the Barometer and Thermometer, whereby Gentlemen and Ladies at the ſame time they Dreſs themſelves, may accommodate their Habit to the Weather, a Performance not only Curious, but alſo Profitable and Pleaſant.

THE ſame Artiſt has made very compleatly in one Frame, a Mercurial and a Marine Barometer, and Thermometer, the Invention of Dr. *Hook*; which performs the ſame by Sea as the Mercurial one doth by Land, being ſo ordered that the motion of the Ship doth not make it Vibrate, as the Mercurial one doth.

Theſe Barometers will be of particular Uſe for Country Gentlemen but more eſpecially to Farmers, in aſſiſting of them in electing Time's, when to Sow or Reap.

So will the Marine Barometers to the Skillful Sailor, in Fore ſeeing Storms at Sea, and their Abatement; and accordingly were of great Service to Dr. *Halley*, Fellow of the Royal Society, in his late Voyage to the South Seas.

Of theſe Weather-Glaſſes, beſides the Common one, there has been of late years ſeveral other curious Improvements.

All theſe are ſo contriv'd, that they may be ſafely ſent to any Place compleatly fixed. And Printed Inſtructions are given with every Glaſs, to explain their Uſes to the meaneſt Capacity.

The Figures of each Glaſs is above.

8 *Trade card of John Patrick*

the opposite side, which made it an attractive and balanced arrangement. Fig 9 shows a later one by Watkins & Smith, of London, and dates from about 1763, when the partnership between Francis Watkins and Addison Smith was formed. A perpetual calendar takes the place of the mirror; this particular calendar begins in 1753, the year following the introduction of the Gregorian Calendar in Great Britain. In addition to showing Easter Day, the Epoch and Dominical letter, it was possible to obtain the time of high water at London Bridge, sunrise and sunset for London and the phases of the Moon. This was done by rotating the knobs on either side of the calendar which operated simple gears at the back of the frame. Below the angled arm of the barometer, which gave a scale magnification of seven, is an oat-beard hygrometer. The barometer has a closed cistern with a leather base and adjusting screw, the spindle of which can be seen in the photograph.

The more general form of the diagonal barometer is shown in Fig 10, it was made by Finney of Liverpool about 1770, and gives a scale magnification of six. The form of the cistern is unusual, and the cover has been displaced to show it. Instruments with this type of cistern were called bottle barometers and, according to de Luc, were common on the Continent about this time.

A number of other makers produced diagonal barometers during the 18th century, in particular Charles Orme of Ashby de la Zouche, who specialised in this type during the second quarter of the century. He also introduced the multiple tube form with not one but two or three tubes, angled at different heights above the cistern, thus gaining greater magnification without making the angled arm too long. Orme was probably the first maker in the country to distil the mercury before pouring it into the tube, then with the tube inverted he heated it so that the mercury boiled and thereby dissolved air was removed.

At the end of the century Balthazar Knie of Edinburgh made diagonal barometers of a more compact form, and also models of the siphon tube type, an example of which can be seen in the Royal Scottish Museum. A few diagonal barometers were produced during the 19th century by other makers, including Samuel Lainton and Charles Howarth, both of Halifax.

Standard barometers

Early in the 19th century there began the separation between barometers for domestic use and those for scientific purposes, and we find barometers being made for the public by instrument makers, but to the designs of the scientific men of the period. From 1774 the barometric pressure had been recorded twice daily at the rooms of the Royal Society in Somerset House, using the barometer by J Ramsden illustrated in Fig 2.

From January 1823 a new barometer was introduced, designed by a scientist and made by a leading instrument maker. It was considered by Francis Baily, FRS, as a 'standard barometer'. It was made by J Newman

9 *Diagonal barometer with calendar by Watkins & Smith c1770*

in 1822 to the directions of Dr Daniell, who also designed the hygrometer and the primary cell, and in describing it Daniell stated that though many barometers of the period appeared to have refinements they were still unreliable because insufficient care was taken to remove all the air from the mercury. The barometer is illustrated in Fig 11a, but part of the case

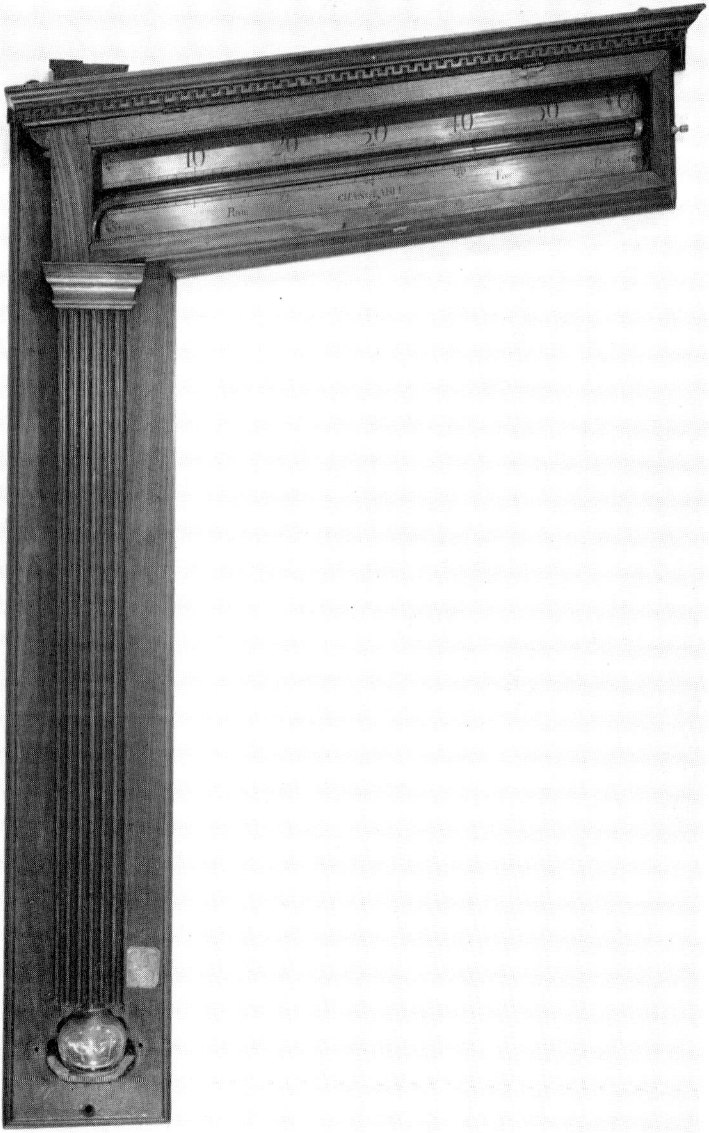

10 *Diagonal barometer by Finney, Liverpool c1770*

around the upper part of the tube is now missing. Francis Baily, referring to it in 1837 stated: '. . . appeals are frequently made to the barometer of this Society, as a standard, by persons engaged in important researches . . .'. The cistern is a large, well-seasoned mahogany bowl, with the thermometer bulb dipping into the mercury; there is also an ivory float with an index mark to indicate changes in the levels of the mercury in the cistern. Allowance could be made for change in the level in the mercury in the cistern with change in the mercury height in the tube. This was known as 'Correction for Capacities' and its values was 1/100. The zero for this correction was when the height reading was 77·663 cms, which was called the 'Neutral Point'. The value of the depression of the meniscus, due to capillary effect, had also been ascertained, and this was + 0·0152 cms to the height. These figures are all engraved on the plate on the front of the barometer case. There was another plate above this on which was engraved a table giving the correction to be applied to the barometer reading to reduce its value to a temperature of 32°F, allowance being made for the expansion of the mercury and the glass at the temperature as shown by the thermometer.

In 1837 another Royal Society standard barometer was made by J Newman, this time to the design of Francis Baily, (Fig 11b). Its most noticeable feature is the two tubes with the common cistern. Each tube could be considered as an independent barometer. One is made of flint glass and the other of crown glass, and the reason for this was to discover, with time, whether one glass had any greater chemical effect on the mercury than the other. The scale is attached to a brass rod which passes through the framework between the tubes, and attached to the lower end of the rod is an ivory pointer. By means of a rack and pinion this ivory pointer can be brought down so as just to touch the mercury surface in the cistern. This was done by rotating the knob at the side of the cistern, and the reading microscope enabled the position of contact to be made more precisely. As the scale was graduated from the tip of the ivory pointer no capacity correction was required. Originally there was a thermometer to the left of the left-hand barometer the bulb of which dipped into the mercury, but when the barometer was repaired in 1877 this was replaced by the present thermometer. This barometer remained the standard until 1856.

In 1842 The British Association for the Advancement of Science took over the old Observatory at Kew (in the Old Deer Park) and set up an establishment for the purpose of designing, improving and testing scientific instruments against standard instruments for meteorological and geophysical purposes. In 1855 a standard barometer was designed by the Superintendent, John Welsh, FRS, which then became the accepted standard barometer, (Fig 12a). It will be seen that there is no scale for measuring the height of the mercury, and the reason for this is that it was the first standard where the distance between the levels of the mercury

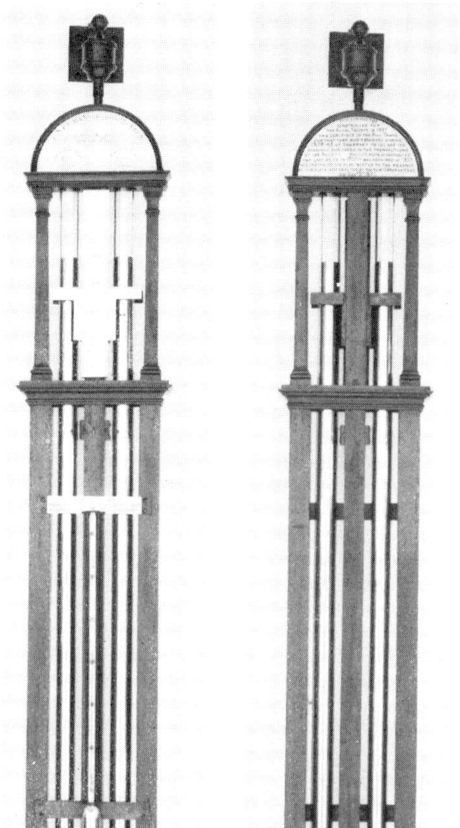

Royal Society standards

11a ABOVE *designed by Daniell, made by J Newman, 1822.* 11b RIGHT *front and rear views designed by F Baily, made by J Newman, 1837*

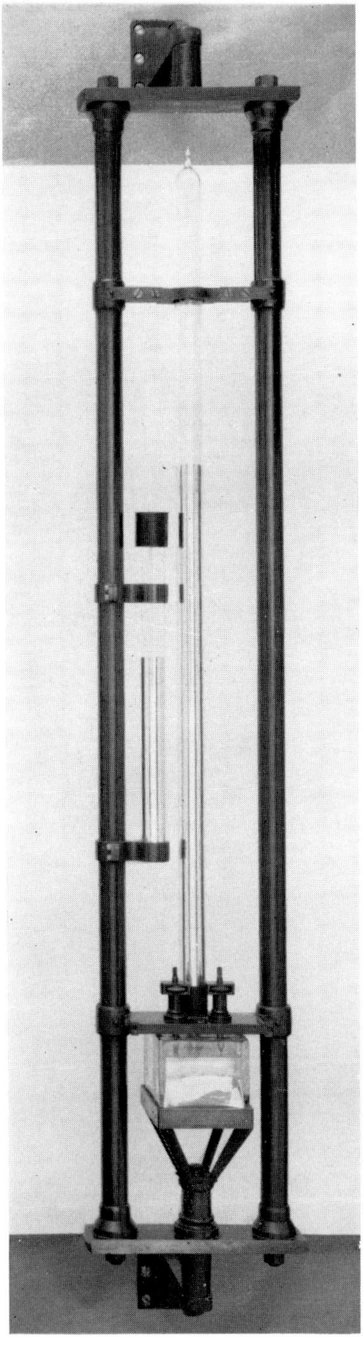

12a *Kew Standard, designed by J Welsh 1855*

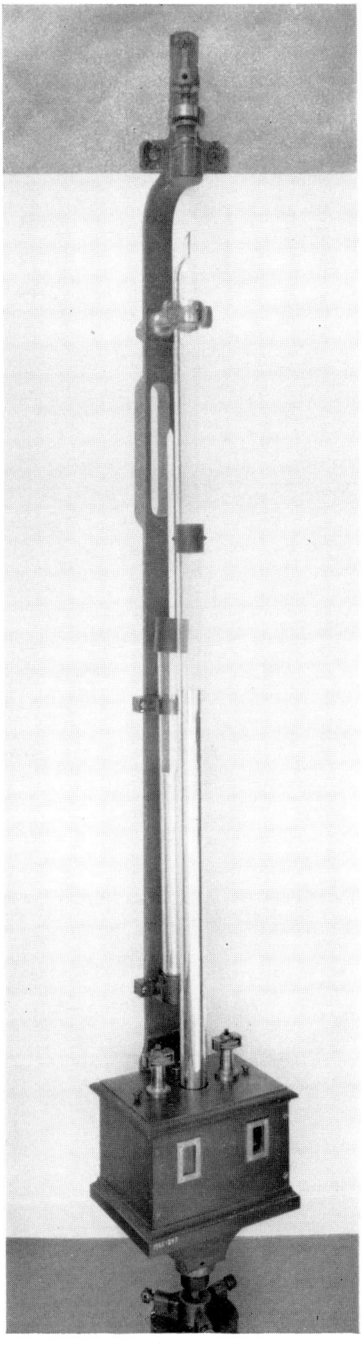

12b *Kew Standard 1860*

in the cistern and the tube was measured with a cathetometer. The steel rod passing through the lid of the cistern had a conical point which was brought down to the mercury surface by rotating the milled-edge screw. There was a line on the upper part of this steel rod which was a known distance from the point, and cathetometer readings were measured from this mark. Initially the temperature of the barometer was that of the cistern and a thermometer was fitted through the lid of the cistern. Subsequently this was replaced by a thermometer in a tube of mercury alongside the barometer tube.

In 1860 this standard was replaced by another (Fig 12b) because it was feared that the former 'might have been injured by the workmen who sometime since were repairing the Observatory'. It was found that there was excellent agreement between these two instruments.

These Kew standard barometers were called 'primary standard barometers'. When an extensive series of comparisons was being carried out, it was too troublesome to use them directly, so a secondary standard was used for the routine comparisons, and this was compared at regular intervals against the primary standard.

The National Physical Laboratory came into existence in January 1900 and took over the testing of barometers, using these same standards. The work continued at Kew Observatory until 1912 when it was transferred to Bushey House, Teddington. The primary standards made in 1855 and 1860 remained in use until 1933 when Sears and Clark produced a new primary standard in which the tube was bored out of a block of stainless steel, with a window of optically flat glass through which the mercury level was viewed. A modified form of this barometer became the primary standard in 1957.

Domestic barometers of the 18th and 19th centuries

The steady increase in the prosperity of the middle classes in Britain, arising from the Industrial Revolution, was reflected in the furniture of the home, and here the mercury barometer was a popular, and often essential item. At about the same time there was a steady flow of Italian immigrants from Holland and France who specialized in glass-blowing. They settled first in London, but later spread out into the country. Many made barometers, though probably only the tubes, employing local cabinet makers to supply the support frames. The quality of the barometers varied considerably. Some makers also produced mirrors and glass vases, others developed into leading instrument-makers providing a wide range of meteorological and scientific instruments capable of great accuracy.

Although the stick barometer, having a fixed cistern of boxwood with a leather base, continued to be made, the siphon type in its several forms now became more popular. During the first two decades of the 19th century the bulb form was popular. Fig 13 shows a barometer made by

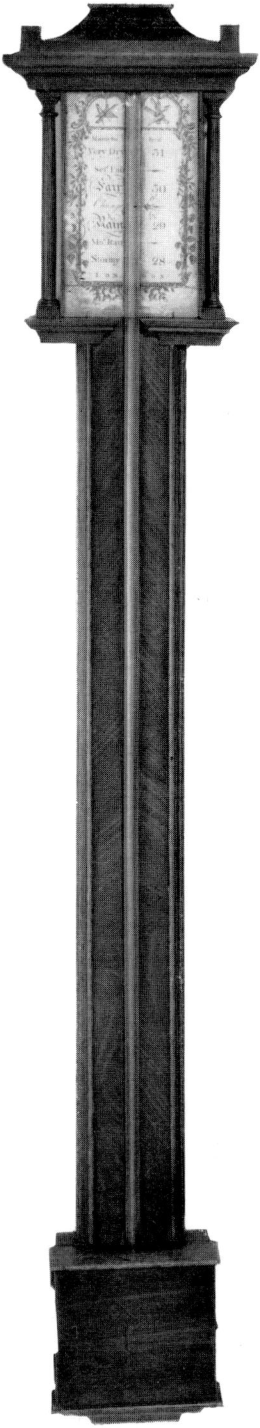

13 *Siphon barometer by D Manticha c1800*

D Manticha, dating from about 1800. The scale was printed on paper which was then varnished. It is possible that this scale was initially covered by glass. All these barometers gave weather indications as well as the pressure scale, which was seldom accurate as no allowance was made for the rise of the mercury in the bulb.

Increasingly throughout the century the wheel or 'banjo' barometer became the most popular domestic form. The advantage was that it was possible to have an enlarged scale so that changes in pressure could more easily be noticed (Fig 3). There was no bulb in the short arm but a straight tube with a float attached to a cord which passed over a pulley wheel to counterpoise weight in a similar tube. A change in the height of the mercury caused the float to move, and so the pointer which was attached to the pulley wheel. The greater the length of the pointer the larger the dial and so the scale on this dial was more open. Clearly it was essential for the barometer to be hung vertically or the float system would be likely to jam. For this reason the outer frame usually carried a spirit level, placed below the dial. At the top of the frame, below the pediment, an oat-beard hygrometer was often inset to indicate the humidity. This form of hygrometer had first been made by Robert Hooke in 1663, and consisted of a beard of wild oats which curled and uncurled with changes in the moisture of the air. It had one end glued to the base, and to the other was glued a straw pointer which moved over a dial engraved with the words *moist* or *damp* and *dry*. An example of this type of hygrometer can be seen below the inclined part of the tube in Fig 9.

FitzRoy barometer

The Meteorological Department of the Board of Trade was formed in 1855 with Vice-Admiral Robert FitzRoy as its first Superintendent. FitzRoy had already produced empirical rules for forecasting the weather, depending not only on the height of the barometer but also on how it was changing and the direction of the wind. These became known as *FitzRoy's Rules* and were published in his *Weather Book* in 1863. At the same time FitzRoy was anxious for fishing ports to be supplied with barometers, and this led him to devise his barometer for the house. It was a siphon type with a small bowl in the short limb which could be covered to keep out the dust. There were two pointers to indicate the barometer readings on two successive days, so that the change of pressure could be readily observed (Fig 14a). His rules were printed on paper affixed to the backboard of the barometer. In addition, there was a Storm Glass, or Camphor Glass, as FitzRoy called it. Such glasses should not be confused with Weather Glasses, described later. According to FitzRoy these were introduced before about 1760 and were sold on Old London Bridge. One of the earliest descriptions of it appears in the *Magazine of Science* in 1839, where it is called a Weather Glass. The contents of the Storm Glass were as follows:

14a *Fitzroy barometer 1858*

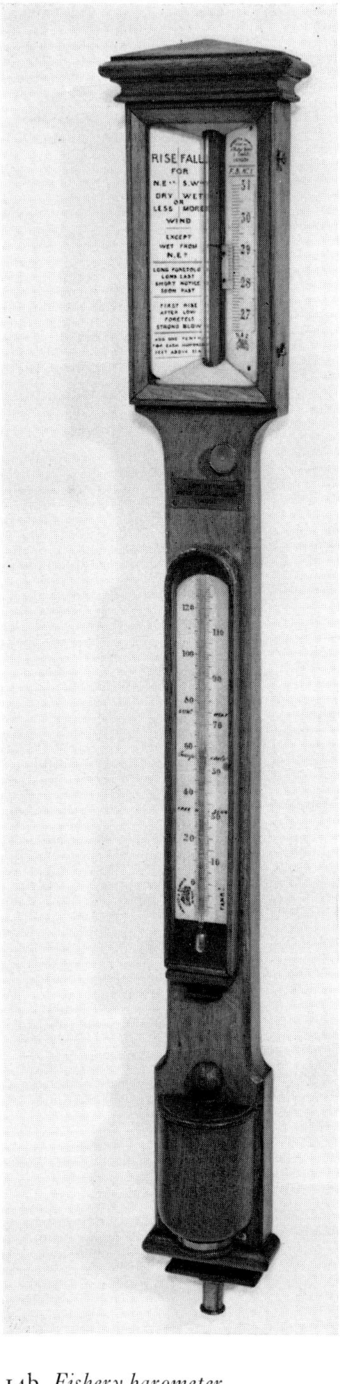

14b *Fishery barometer*

2 grams	Camphor
0·5 grams	Potassium nitrate
0·5 grams	Ammonium chloride
28·5 ml	Alcohol
28·5 ml	distilled water

The solids were first crushed and the camphor dissolved in the alcohol, while the powdered salts were dissolved in the water. The two solutions were mixed in a tube, about 15 cms long and 1·6 cm diameter, which was then corked and hermetically sealed by covering with sealing wax. Initially they were regarded as curiosities because changes occurred in the liquid when there was no change in the height of the barometer. According to FitzRoy these changes depended upon the wind direction, but not its force, and the electrical tension in the atmosphere. In fine weather the solid matter settled as a white precipitate at the bottom of the tube; with the approach of rain the precipitate filled the tube in the form of small crystals floating in a clear liquid, but 24 hours before a storm or very high winds the liquid became turbid throughout. We know now that these storm glasses have no practical value in foretelling the approaching weather. The supporting frame of many of these barometers was of oak. Although they are advertised in the catalogues of practically all the makers over the period 1865 to 1914, few seem to bear a maker's mark.

Fishery or sea coast barometers

By 1858, FitzRoy had arranged for special barometers to be issued by the Board of Trade to fishing ports, Coast Guard and RNLI Stations in order that they could make use of his rules, empirical at that time, in giving warning to sailors of the approach of storms. These were known as fishery or sea-coast barometers. Their main features can be seen in the one illustrated in Fig 14b, which was the second one ever to be issued, to St Ives Coast Guard Station. The tube was a strong one of wide bore and of the closed cistern type, and the scales were of white porcelain with clear markings so that they could be easily read when placed in a public building or lifeboat station. There were usually two verniers, one for the day's reading and the other set on the previous day. The whole was enclosed in a strong oak case with a glass door.

It was not until 1861 that the first storm signals were displayed at British ports on instructions from the Meteorological Department in London, but these signals could only be displayed at the larger ports which were connected by electric telegraph with London. They were discontinued soon after FitzRoy's death in 1865, but were re-introduced in January 1868 as the result of public demand.

Marine barometers

The ordinary barometer was unsuitable for use at sea because the motion of the ship caused the mercury to pump or oscillate in the tube, making accurate readings of the true height virtually impossible. Robert Hooke made various suggestions to overcome this in a paper to the Royal Society in January 1668. However, the first barometer designed specifically for use at sea was produced by Edward Nairne about 1771–72. Captain Cook used one on his second voyage to the Pacific and again on his final voyage. Nairne's marine barometer was of the closed cistern type, made of wood, with a tube of restricted bore or 0·145 cm for all but the top 10 cm, which had a bore of 0·8 cm. This tube was in a wooden case which could be mounted in gimbals, and further to assist it to remain in the perpendicular position a weight was fastened to its bottom. Ever since then all marine barometers have had tubes with restricted bores.

Throughout the first half of the 19th century an increasing number of marine barometers were made and all were similar, of the closed cistern type with a leather base below which was the adjusting screw. This screw was covered by a screw-on cap so that it would not be exposed to view, and thus reduced the risk of interference by unauthorised persons. The tube was of the constricted form as invented by Nairne, and the whole was encased in a wooden frame. There were holes in the sides of this frame to enable the barometer to be supported in gimbals. Fig 15a shows one typical of its time, made by R B Bate of London late in the period.

In about 1885 the design of marine barometers was investigated by John Welsh of Kew Observatory. As a result, 'Kew pattern' was developed by him and made by Patrick Adie, the London instrument-maker The tube had, in addition to the constricted bore as in earlier marine barometers, a Bunten air trap immediately below the constricted part of the tube. This trap, in the form of a pipette, had been invented by Bunten of Paris in 1824, and it prevented any air which may inadvertently have entered the tube passing up to the top of the tube above the mercury and so affecting the vacuum, which it was necessary to prevent if the barometer was to operate properly. The cistern of these marine barometers is a closed one of iron, without any adjustment, closed completely except for a small protected aperture open to the atmosphere. The whole is in a brass frame mounted in gimbals. The scale is calibrated to allow for the varying level of the mercury in the cistern with change of pressure.

There has been practically no change in the form of this marine barometer since 1856 up to the present day, except that steel has replaced iron for the cistern. Fig 15b shows an early one by Patrick Adie.

When the Board of Trade issued barometers to fishing ports and Coast Guard stations, they also issued them to ships' captains who were prepared to take regular meteorological observations when at sea and send in their meteorological log books to the Department. A complete record of the movements of the marine barometer illustrated has survived. It was issued in February 1858 and was used in various ships until November 1889, and passed to the Science Museum in 1892.

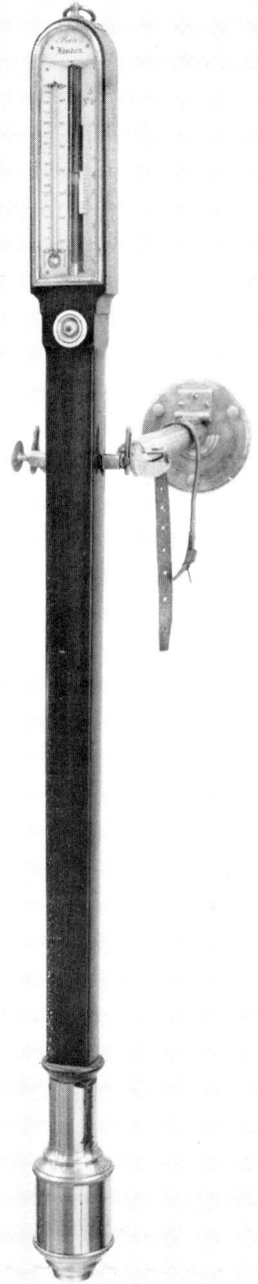

15a *Marine barometer by R B Bate c1840*

15b *Marine barometer by Patrick Adie 1857*

Station barometers

These are used at meteorological reporting stations and are of the Kew pattern. They are almost identical with the above, but with one difference: the constricted bore of the tube is not so fine. Up to 1914 the scale was calibrated in inches or centimetres of mercury, but from 1st January 1914 the scale was calibrated in millibars.

Fortin barometer

Designed by Nicolas Fortin of Paris in about 1808 it is probably the best known form of the mercury barometer, and it is found in practically all college science laboratories to this day. The closed cistern has a leather base, and there is a glass collar, held tight by leather gaskets, forming the upper part of the cistern so that the mercury level can be seen.

Mountain barometers

These are portable and used by travellers, explorers, and sometimes surveyors, for determining the height of a mountain. As far back as 1648 Perier, brother-in-law of the famous Blaise Pascal, took a barometer up the Puy-de-Dome, leaving a similar one at the base in Clermont with an observer, and found that the height of the mercury in the tube decreased as the mountain was ascended, while there was no change in the pressure at the base of the mountain. It was many years later that a mathematical relationship was found between the decrease in pressure with height in the atmosphere. In 1755 J A de Luc devised his portable barometer for mountain work. This was a siphon-type in which it was possible to close the short arm by means of an ivory stopcock, after the barometer had been tilted so that mercury filled the closed tube. This prevented air entering the tube and also prevented the mercury striking the end of the tube while it was being transported, lessening the risk of breaking the tube. In addition, he adopted an unusual method of graduating up and down from a zero near the top of the shorter limb so that, to obtain the height of the mercury, these two readings were added to give the height of the barometer. Fig 16 shows a de Luc mountain barometer made by Nairne and Blunt of London in about 1785. The plumb-bob in the travelling case enabled the instrument to be placed vertically before the reading was taken, while the right-hand thermometer could be removed from the case and then used to measure the ambient temperature, while the other registered the temperature of the barometer.

Because the pressure decreases with height the range of the scale of mountain barometers is much greater than that for the normal barometer used at low altitudes. The scale is usually graduated down to about 53 cm of mercury, corresponding to the pressure around 2700 m to 3050 m.

Various forms of mountain barometers were developed throughout the 19th century, and it was usual to suspend them from a tripod when the pressure reading was required.

Even for meteorological reporting stations at high altitudes it was necessary to modify the normal station barometer. The Kew-pattern long-range barometer, used today at such stations, is an example, and the scale is calibrated from 100 mbs down to 700 mbs.

Double and folded barometers

The double barometer was invented by Hooke in 1668, but subsequently others, including Christian Huygens, modified it. These were essentially siphon barometers with a light liquid above the mercury in the open limb which was therefore much longer than normal. This liquid was initially water, but later other liquids such as oil of tartar and alcohol were used. The scale was against the open limb and was therefore inverted.

The disadvantages of this type were that the liquid slowly evaporated and its volume changed appreciably with change of temperature. More seriously, the tube became very dirty in the region where the liquid moved with change of pressure.

Various modifications were made to this type. Some had two liquids in addition to the mercury. The majority were made in the 18th century. One form, made particularly towards the end of the 18th and the beginning of the 19th century by Italian glass-blowers in Holland, was called *barometer en contraleur*. With these there was an ordinary barometer with a double barometer alongside it on the same frame. Presumably, the ordinary barometer gave the pressure value while the double barometer showed the daily change more clearly on the expanded scale.

The folded barometer was invented by Amontons in 1688, and resulted in the instrument being smaller and so more easily portable, while at the same time the scale was enlarged so making it easier to note the changes in pressure. The mercury column was split into two lengths with a lighter liquid between them, and the same liquid was above the mercury in the open limb. There are two of these in the Science Museum, one by D Manticha and dated 1781 but now no longer containing either mercury or liquid. The other one is by Joseph Betali of Paris, and is illustrated in Fig 17. Folded barometers were made during the 18th century, but today few exist.

Air barometers

These are instruments, primarily intended for marine use, in which a gas occupies the vacuum space in the ordinary mercurial barometer. Since the volume of the gas changes with change of temperature as well as change of pressure it is necessary to measure this temperature in order to obtain the pressure.

The first instrument of this type was described to the Royal Society by Robert Hooke in January 1668. It was described some years later by Edmund Halley, who stated that he had one on his voyage to the South Atlantic at the turn of the century, 'and it never failed to prognostick and

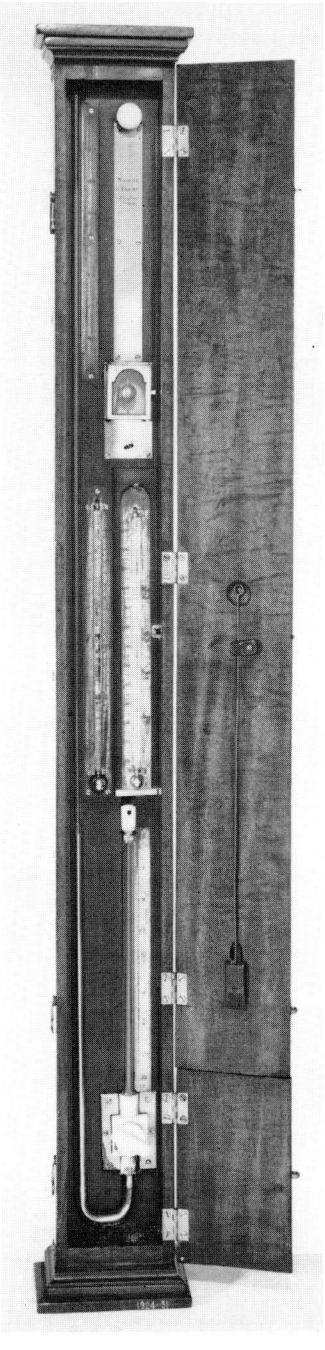

16 *de Luc Mountain barometer by Nairne & Blunt c 1785*

17 *Folded barometer by J Betali of Paris*

give early notice of all the bad weather we had, so that I depended thereon, and made provision accordingly'. In 1705 Guillaume Amontons of Paris described an improved type which he called, 'A Barometer with mercury for use at sea'.

The sympiesometer is the best known and the most widely used example of an air barometer. It was invented by Alexander Adie, the Edinburgh instrument-maker, in 1818, who described it as 'An Improvement on the Air Barometer, which Instrument is to be called a Sympiesometer'. Fig 18, shows an example by L Casella, made early in the second half of the century; these instruments were used at sea until near the end of the century. The upper bulb was sealed and contained hydrogen above the liquid which was coloured almond oil. The lower bulb above the liquid level was open to the atmosphere. The instrument had first to be calibrated against a barometer. The pressure scale moved up and down a slide, and to obtain the pressure the index was set against the ambient temperature on the right-hand temperature scale. The pressure was then obtained from the pressure scale corresponding to the top of the liquid in the tube.

Various other forms of the air barometer were invented during the 19th century. In 1839 Ronketti modified Adie's sympiesometer by placing the thermometer bulb in the gas in the lower bulb, which was sealed off while the upper bulb remained open to the atmosphere. He also replaced the almond oil by mercury.

Another form invented by T T MacNeill of London in 1870 is sometimes described as a Boylean-Mariotte barometer because it is a direct application of Boyle's Law, (Fig 19). This is a very compact form of air barometer, about 35·5 cm long, and it could register pressures between about 33 cm and 82·5 cm of mercury, and so was suitable for use at high altitudes. Here, the tube was open to the atmosphere and the cistern was a leather bag with a screw and piston head below it. There was also an air chamber around the lower part of the tube open to the atmosphere when the tap was open. To operate this barometer the tap was closed, enclosing air in the chamber at atmospheric pressure. The screw below was steadily raised so that mercury rose in the chamber and the tube. When a mercury bubble began to appear through a diaphragm in the chamber, as seen through the glass window, then the air in the chamber had been compressed to a fixed volume. The reading of the mercury level in the tube against the scale gave the atmospheric pressure. The diagram indicates the essential features.

The aneroid barometer

The real departure from the principle of the Torricellian mercury barometer for measuring atmospheric pressure was the aneroid barometer, which is now the only portable form in general use. The aneroid barometer was invented by Lucien Vidie (or Vidi), of Nantes, in 1845. It has

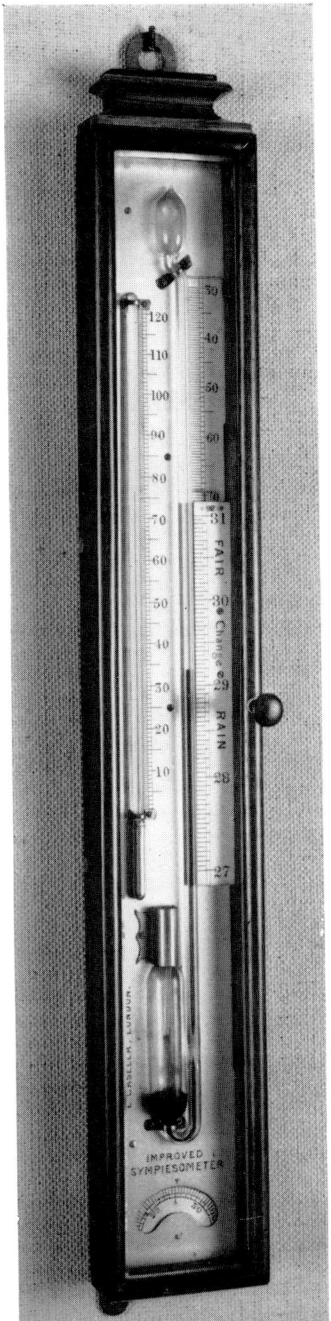

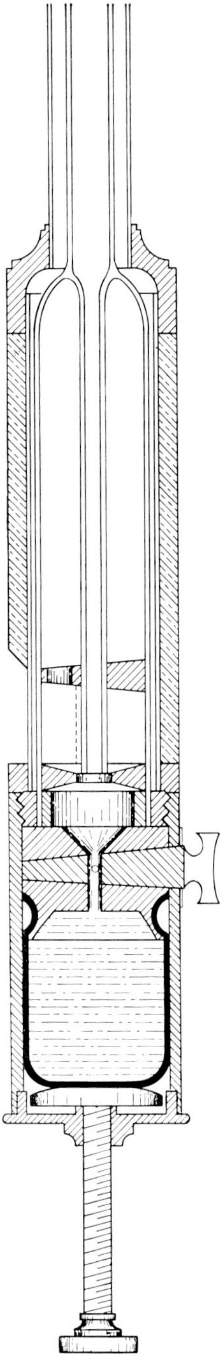

18 *Sympiesometer by L P Casella* 19 *Boylean-Marriotte Air Barometer*

remained essentially the same to the present day. Even before 1654 Blaise Pascal had suggested the construction of a metallic barometer without mercury, whose office the bellows performs which the weight of the air tries to compress against the resistance of the steel spring. In 1797 N J Conte, of Meudon in France, suggested a similar form of barometer, but it was Vidie who finally invented and constructed the aneroid barometer. After many tests he brought out his first patent in 1844, but not in his own name. The form which is essentially as we know it today was patented in the following year.

The aneroid barometer consisted of a flat cylindrical metal box with the circular surfaces of thin metal corrugated in concentric circles. This box was almost completely evacuated, containing a little gas at very low pressure. The residual gas was to compensate for temperature change. At the centre of the corrugated surface was a smaller pillar to which was attached the principal lever system which operated the pointer. Changes in the barometric pressure caused the metallic membrane to move up or down and this movement was magnified by the lever system. There was a leaf-spring to hold the membrane in its normal position as if the box had contained air at normal pressure. The purpose of this was to prevent the total collapse of the evacuated box.

This barometer first appeared in England in 1847, sold by E J Dent. It is interesting to note that the early tests on these aneroids were made to observe their variation with changes of altitude, and small changes at that, by noting the change with the gradient of a railway line.

The aneroid barometer rapidly became popular in the home in England, and was sometimes substituted for, or adapted to, the form of the older wheel barometer, or 'disguised' as a mantel clock.

At about the same time a partial mathematical relationship was developed for the change of pressure with altitude, and soon instruments were available with a bezel, graduated in altitude, which could be rotated. The zero of this scale was set against the pressure shown by the pointer before any ascent was made. As the ascent was made the pressure fell and the pointer not only indicated the pressure but the altitude. Naturally it had to be assumed that there had been no change in the surface pressure during the ascent. Such aneroid barometers soon became the essential equipment of travellers, explorers, surveyors and balloonists, and some very attractive aneroids with a height scale bezel were produced in a miniature form, about the size of a pocket watch. During this century the aneroid capsule has formed the sensitive element of aircraft altimeters and machmeters.

The fact that the sensitive membrane had to drive the central pillar, together with the lever system, meant that it had to overcome friction at various points. It also meant that the metal was subject to fatigue effects and thus changed its characteristics. In 1922 G Paulin of Stockholm improved the aneroid by arranging that the metallic diaphragm could only move over a very small distance and was always in the same position when the reading was taken, whatever the actual pressure. The reading pointer was rotated so that a 'tendency' pointer was brought to zero, indicating that the membrane had been returned to its normal position. It was claimed that the elastic hysteresis of the metal was reduced to practically nil. Such barometers were particularly useful for indicating height.

In 1960 a precision aneroid barometer was patented by J F Grandidge, which in a modified form is now rapidly replacing the mercurial barometer at sea. In the later version the metallic membrane does not have to do any external work to drive the lever and pointer system, and the friction of that system has been overcome. There is a separate contact made with the membrane which is moved by a micrometer screw, and this also operates the digital counter, so giving a direct reading of the pressure. The moment of contact between the point of the micrometer screw and the pressure capsule is indicated by means of a simple electronic system in which there is a small cathode-ray indicator. This precision aneroid is illustrated in Fig 20. A portion of the casing has been replaced by Perspex so that the capsule and counter can be seen.

Returning to the year 1846, another form of the aneroid was invented by a German engineer, but he did not patent it until 1849, the same year that E Bourdon of Paris patented an almost identical one. In this the sensitive element was a flattened tube of elliptical cross-section, curved round in a circle and evacuated. With change of pressure the tube curled and uncurled slightly, and this change was magnified by a simple lever and

20 *Precision Aneroid barometer 1965*

pointer system. Although the Bourdon type was more popular initially in France than the Vidie type it was not so reliable and soon fell out of favour. However, the Bourdon tube is still used in pressure gauges, but in these the steam pressure is applied to the inside of the tube. There is an example of the Bourdon type aneroid barometer in the Science Museum, and its simple construction can be clearly seen.

Barographs

The barograph is an instrument which gives a recording of the barometric pressure. Thus it is valuable in showing how the pressure changes, especially the sharp changes associated with thunderstorms and 'cold fronts'.

The earliest suggestion of a barograph was put forward by Sir Christopher Wren in his comprehensive 'weather clock' devised in 1663. It was added to by Hooke, but only a brief description of it exists today. It seems that the readings were not continuous, for the marks were made at regular intervals.

The earliest barographs actually made were those by the clockmaker Alexander Cumming. These were barograph clocks. The first, made for George III in 1765, is now in Buckingham Palace. Cumming made two more, one of which was for himself, in 1766, and after his death it was purchased by Luke Howard, regarded as the father of modern meteorology, who gave the names to the clouds which we still use today. This barograph was operated by the float of a double siphon barometer. There were two tubes concealed in the two corinthian columns (Fig 21) which dipped into the common cistern. The upper portions of the tubes had a much greater diameter than the tubes themselves so that the maximum change occurred in the level of the mercury in the lower limb. The pen arm was attached to the float in this short tube and gave a continuous trace on an annular chart around the clock face, the chart being driven by the clock. This barograph was set working again for the bicentenary of the birth of Luke Howard in November 1972, and is now the oldest working barograph in the world. It is on loan to the Science Museum for ten years (from 1973).

The main problems to be overcome with recording instruments were the effects of friction, particularly of the recording pencil over the paper. This was not so apparent in the Cumming Barograph Clock as the daily movement is only about 3 mm on the average. The recording on later barographs was of an intermittent nature, the pencil being normally clear of the paper, but was struck at regular intervals so as to make a mark on the paper. In 1780 Changeux produced a barograph similar to Cumming's, but with this intermittent form of recording.

In the middle of the 19th century a number of mercury barographs were developed which gave weekly or daily traces. These showed far more clearly the changes in the barometric pressure than was possible with the

21 *Barograph Clock by Alexander Cumming 1766*

Cumming barograph. There are several such mercury barographs in the Science Museum which show how these developed, and they include examples by K Kreil of Prague, 1841; M Pillischer, 1850; Admiral Sir A Milne, 1857; L Redier of Paris, 1874; and W H Dines, 1904, the last two of which gave continuous recordings. In 1841 Kreil devised an instrument in which the clock drew a frame carrying the chart horizontally at a uniform rate, while the float in the short arm of a siphon barometer was connected by a string which passed over a sector. To this was fitted an arm supporting the pencil, which was struck by a hammer operated by a capstan on the clock face.

An unusual barograph, first proposed in 1854, was developed by an engineer, Alfred King, of Liverpool, who described it as a 'floating barometer'. The barometer tube, made of iron, was suspended over a large wheel to a counterpoise weight, while its open end dipped into a large iron cistern. The recording pencil was carried on a frame on the counterpoise arm, while the chart, which ran for two days, was fitted to a vertical drum which was clock-work driven. There was an iron float around the tube where it entered the cistern, and this served to control the movement. The system was initially balanced at about 73·6 cm barometric pressure. With increase of pressure the equilibrium was disturbed and the tube would descend, but the float would displace mercury in the cistern and eventually equilibrium would again be restored. The magnification of the movement, and so the barograph reading, depended upon the relative diameters of the tube and the float. In this barograph, which was installed at Liverpool Observatory in 1862 (Fig 22), the magnification was $5\frac{1}{2}$. Such striking barographs of the industrial age were installed in a number of observatories, and it is understood that there is only one in operation today, at Edgbaston Observatory.

Following the work of Fox Talbot on his 'photogenic drawings' in 1839, the beginning of photography, the first photographic recording of barometric pressure was obtained by T B Jordan in the same year. Although photographic recording had certain disadvantages, its great advantage was that any errors on account of friction were eliminated. Sir Francis Ronalds, the first Superintendent of Kew Observatory, developed his apparatus for the photographic recording of barometric pressure as early as 1845. This 'photobarometrograph' was steadily improved until by 1867 there were examples at other observatories. A cistern barometer was used mounted to incorporate temperature compensation.

A light beam was focused on the upper part of the tube in the region of the meniscus, with the photographic paper in front of the tube, on a clock-work-driven drum. The trace showed one portion black and the other white, the pressure reading corresponding to the boundary between the two. The modified Ronalds version and a Kew photobarometrograph are exhibited in the Science Museum, while the original one at Kew operated from 1867 until about 1969.

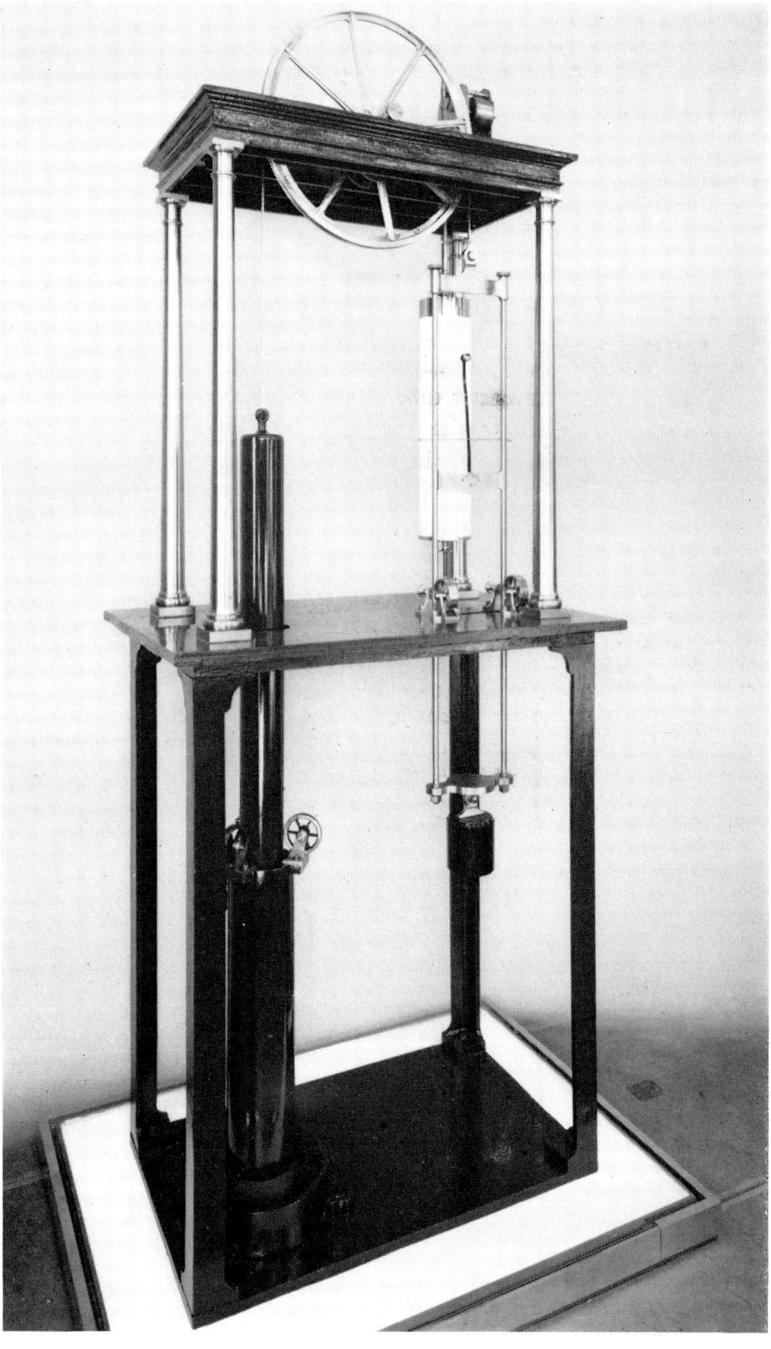

22 *King Barograph 1862*

The familiar barograph seen today in homes is of the aneroid type. In these a number of aneroid capsules placed above each other, form the sensitive element and operate the pen arm through a lever system.

Water and glycerine barometers

The mercury barometer developed from the study of the problem of raising water in mines. In the course of this study what amounted to water barometers were constructed in order to discover what height water could rise in a tube or pipe. In this country the first water barometer was erected by Richard Towneley of Towneley Hall in Lancashire as early as 1661, and two years later the Royal Society had one erected. Thereafter, little interest was taken in them until the beginning of the 19th century. Several new versions were then constructed, some with liquids other than water, as they were not likely to freeze during the winter months. Towneley had one of the siphon type erected against his house with the shortest limb extending into the basement. The tube was mainly of lead piping, with a float in the lower limb to indicate the movement of the liquid. The range was about 53 cm. He found that there was a continual discrepancy between its readings and those of the mercury barometer, due to temperature effects on the liquid and the lead piping, so he discontinued the experiment and dismantled it.

In 1823 Sir John Leslie erected one at Edinburgh University and in 1830 another was erected for the Royal Society by Professor Daniell. This latter was of the cistern type and had a layer of castor oil on the water surface in the cistern to prevent air dissolving in the water. The tube was 12·9 m in length and made of glass tubing in one piece. Similar discrepancies were found with this one as with that made by Luke Howard. When the Royal Society left its rooms in Somerset House this barometer was transferred to the Crystal Palace, but was destroyed by fire in 1866. Probably the principal defect of water barometers was due to the vapour pressure of the water and its change with the change of temperature. In the 1870s T B Jordan developed a similar barometer using glycerine coloured red with rosanaline, instead of water, as the vapour pressure was much lower. The mean height of the barometer was 8·23 m and the scale range about 76 cm. Here was a barometer which could be clearly read by anyone when it was set up on public display. He had one erected in Folkestone Harbour Station to enlighten cross-channel passengers. The Royal Society tested one at Kew Observatory, and *The Times* had one erected in Printing House Square in 1880. In the very week that *The Times* started to report its readings a violent storm struck the British Isles which helped to draw attention to this barometer. One was set up in the Science Museum early in 1877 and continued working until 1923.

There are a few glycerine barometers in operation today, but the only one on public display and in working order is in the Royal Scottish Museum in Edinburgh.

Old Dutch weather glasses

These are glass vessels, with a spout like a coffee pot, containing water which may be coloured. Their height is generally about 28 cms, and their backs flat so that they can be hung conveniently on a wall. They are referred to under various names such as water barometers, weather-glasses, old Dutch weather glasses and Liège barometers. This last name is derived from the fact that many were made in the glass works at Liège. Fig 23 shows a plain water barometer, but those made at Liège were more artistic and illustrated the glass workers' art.

These weather glasses first appeared in this form in the 17th century, and it is probable that they were derived from an earlier version, like that

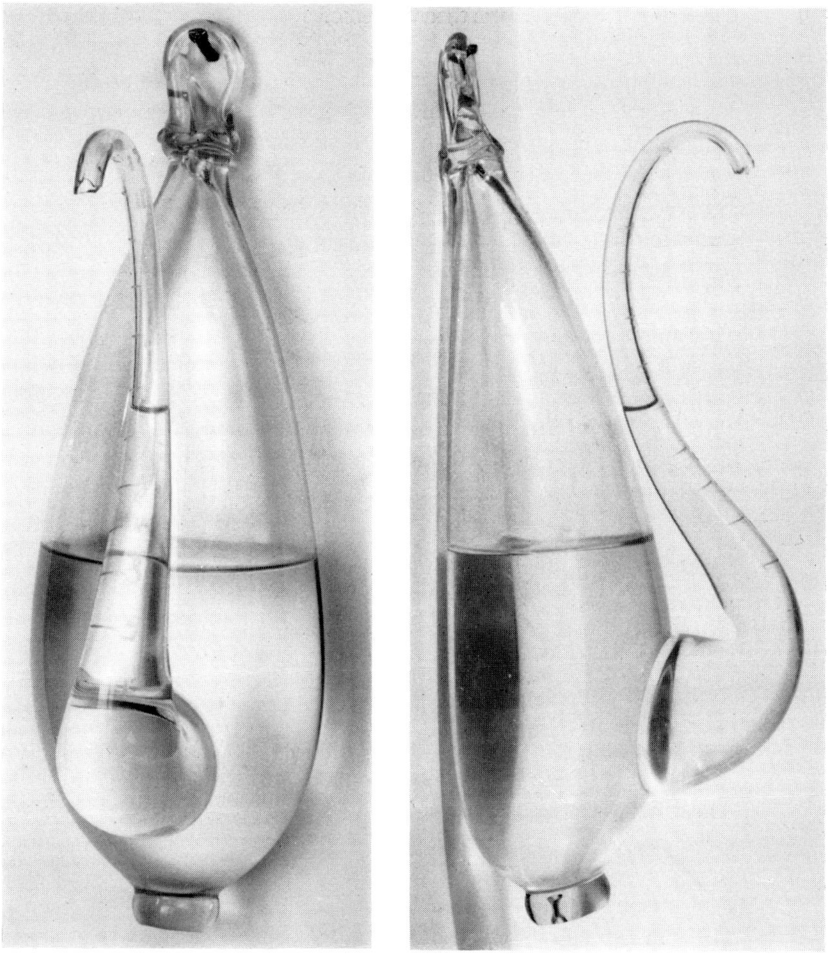

23 *Old Dutch Weather Glass – two views*

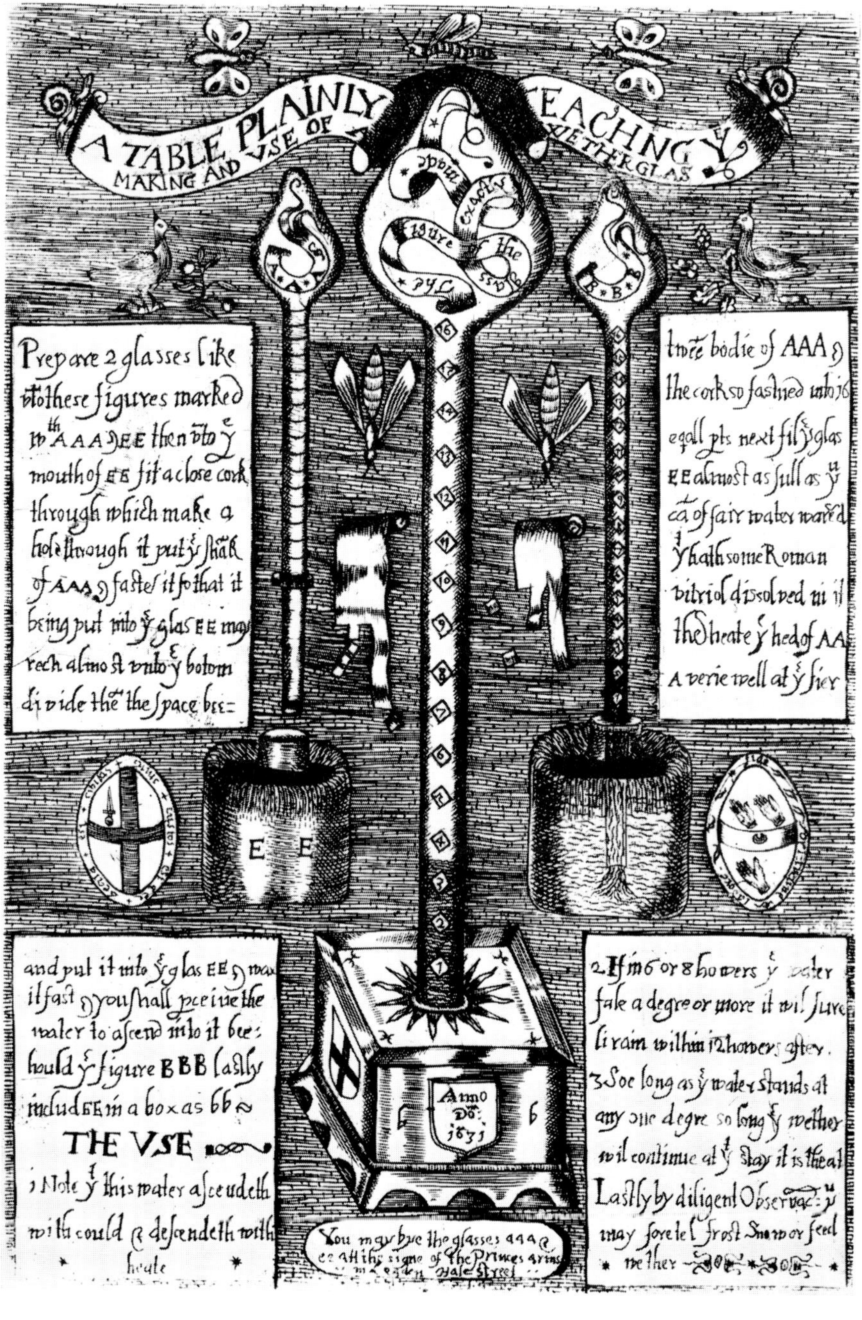

24 *Engraving of Weather Glass 1631*

illustrated in an engraving bearing the date 1631, which states that such weather glasses could be bought at 'the Signe of the Princes Arms in London Hale Street' (Fig 24).

With the increase in atmospheric pressure the water sinks in the spout, and with decrease of pressure it rises. Obviously this change is produced by change of temperature as well as change of pressure, but if the glass is kept in that part of a room where the temperature is reasonably constant, then it does show change of pressure quite adequately.

Such glasses were made in this country until 1939, and quite recently a London store imported some from Switzerland and sold them as Christmas gifts. As late as 1897 a patent was taken out for an improvement in these glasses. This was graduation of the spout so that changes could be observed more clearly. There is a very elegant one, which bears the date 1710, in the Frans Hals Museum in Haarlem, Holland.

Further Reading

The History of the Barometer W E Knowles Middleton, 1964.
English Barometers 1680–1860 Nicholas Goodison, 1977.

Printed in England for Her Majesty's Stationery Office by W. S. Cowell Ltd, Ipswich
Dd 496468 K32 4/78